Science Josh – 2012 Edition

Science Josh – 2012 Edition

Science Josh – 2012 Edition

By Josh Price

Join us at:

www.sciencejosh.com

@sciencejosh

Facebook.com/sciencejosh

Science Josh – 2012 Edition

To all my Twitter followers, blog readers and Facebook likers.

And to Alex Hackman for many interesting inspiring talks.

Science Josh – 2012 Edition

Science Josh – 2012 Edition

Copyright © Joshua O M Price, 2012

All Rights Reserved

Joshua O M Price has asserted his rights under the Copyright, Design and Patents Act, 1988 to be identified as the author of this work

This book is sold subject to the condition that it shall not, by way of trade or otherwise, be lent, sold, hired out, or otherwise circulated without the authors prior consent in any form of binding or cover other than that in which it is published and without similar condition including this condition being imposed on the subsequent purchaser.

All images are property of their owners and creators, used with permission of their owners.

www.sciencejosh.com

Cover design by Josh Price

ISBN 978-1-291-23026-0

Science Josh – 2012 Edition

Contents

July 2012

My First post	13
Debate an Infinite Universe	15
Relative to nothing – Relativity	19
Time Travel: is it possible?	22
Time Travel: Back to the Future	26
Understanding the incomprehensible	29

August 2012

Global warming: Is it true?	32
Global warming: An Introduction	34
Global warming: Carbon Dioxide	37
Genetic Modification: Possibilities and Capabilities	40
Global warming: CFC's	43

September 2012

Is this life real?	45
What makes a human?	48
Inside a black hole	51
Welcome to the Launch	53
Mind Blowing	55
The New Magnesium	58
You are levitating right now	60
Very Interesting Facts	62

October 2012

Is Harry Potter Real?	65
Flying Cars?	68
If a tree falls…	72
The Higgs Boson	75

Science Josh – 2012 Edition

The Higgs – Take 2	79
Dawkins	81
Why do cats always land on their feet?	83

November 2012

Who made God?	85
The Third Law of Thermodynamics	88
Stimuli-Responsive Polymers and Invisibility	91
Competitive Pain – The Science of Pain	94
Origin: The creation of the universe, the laws of science and time	99

December 2012

Why are my headphones tangled, again?	103
Could there be life on other planets?	107
Will the world end this Friday?	114

#2 Will the world end tomorrow? – From the Bible

117

About the Author 121

Science Josh – 2012 Edition

My First Post

July 2012

So, before I start with my actual posts that hopefully contain something interesting, it is probably best if I tell you a little about me!

Well, I am 15 and live in West Sussex. I like computers, drama and debating/arguing with people. I often have mad business ideas and spend a lot of time being 'nerdy'. I have started a blog for two reasons; 1. I am a bit bored and think it will be fun. 2. I want to share my views and opinions on things that are occurring in the world now.

I will babble on about lots of random and probably boring stuff, however, will try and make it interesting. Please feel free to disagree with me as this leads to debate which I love!

Science Josh – 2012 Edition

Debate: An infinite universe

July 2012

Now this is a subject of debate. Today I got into a debate about this:

Is it possible for the universe to expand? If the universe is expanding, which Red shift proves, what is it expanding into?

Let me say it differently; if something (The universe) is expanding into nothing, what is that nothing and where is it coming from? If there is nothing and the something is expanding into it, surely, that nothing must be something.
In other simpler words; will the universe (something) run out of nothingness to expand into?

I argue that it won't run out. Nothing is effectively a lack of everything, there is nothing. So nothing is infinite because there is nothing to limit it. The something can expand into the nothing, giving the nothing matter and the laws of science therefore are in action in that area of now, something. Some of the nothing has been filled with something and so the amount of nothing has diminished. However, because nothing is continuous and infinite nothing has been used up because there never was anything to use up.

In writing this I have confused myself and wonder now if time still affects nothing, but then how could it? Also, what is a vacuum? Is a vacuum nothing? EM waves can travel through a vacuum so there must be something making a vacuum have substance at its centre. Can EM waves travel through nothing? Could light therefore leave the universe and travel into the nothing? Would the nothing just become something? Is that how the universe expands even?

I must now make clear that what I say is not fact or even researched. It simply is me questioning how things work based on my understanding and perception of everything.

Later addition after some research:

The nothing may not exist! Let me explain. We can only prove that the universe is bigger than 26 billion light years. We know that the universe is expanding. However we cannot prove the existence of what it is expanding into because it is so far away. The something (The universe) is expanding into something that has not yet been proved to exist… nothing. This means however that the universe may not be expanding into nothing: the absence of an existence of something, but nothing: the absence of evidence for the existence of something.

Now is the issue of the universe actually expanding. Surely matter cannot be created so the matter that is currently here must therefore be moving apart to allow expansion. If everything is moving apart so are atoms and particles. This would mean that, with a fast rate of expansion that planets and stars would expand to a point that the bonds are not strong enough to hold the planet together. This could mean that gradually all planets and the entire contents of the cosmos would break apart making expansion possible. However, as there is no evidence for this taking place, it most likely isn't. This means the universe isn't expanding. And that makes the whole argument you have just read pointless.

Science Josh – 2012 Edition

Relative to nothing – Relativity

July 2012

I recently read part of a book called Why Does $E=MC^2$? It explained relativity and I thought I would share what I gained from it with you; and add my own thoughts and general babble relating to the matter.

From the earth we don't seem to move, instead everything moves around us. But when you go to sleep in your bed you move location, or do you?

If you go to sleep at 12pm and sleep for 3 hours before waking up. At 3am you get up, go to the toilet and get a drink from the kitchen. This takes you 12 mins. After this you return to your bed at 3:12am. To you, at 3am and 3:12am you are in the same place. But you're not. In those minutes the world has been spinning on its axis, the earth is orbiting the sun, the sun is moving within the galaxy and so on. You are moving. You are moving location in space within those 12 mins even though to you, you have not. In fact you would have moved thousands of miles in those minutes!

But then have you actually moved? You see, you are, at 3am and 3:12am, in the same location relative to your bed and your house. However relative to Mars you have moved because of you being on the earth and the earth orbiting. You have moved location in space relative to Mars. All make sense? I hope so!

Now here's the issue! How do you prove something is moving, not relative to something else? How could you prove that a planet is moving if there are no other planets or stars? You could have a huge space sized 3D grid to measure location against. Good idea? But then that would be relative to something, the grid. And to create such grid you would need something to make the grid relative to.

That is relativity.

Science Josh – 2012 Edition

Time Travel: Is it possible?

Is time travel possible?

This is a very complicated subject, and I will only be looking at one way time travel could be possible. This way relates to spacetime; it is a very complicated thing and I don't really understand it at all. I will try to use my understanding though, to explain if time travel could be possible. To explain what happens and how it could be possible, I will refer to an experiment involving Muon's.

A Muon is almost identical to an electron, but it unlike an electron has a lifetime. It is very unstable so, decomposes to other particle things after 2.2 microseconds. The lifespan is important in this, so try to remember that.

In this experiment there were lots of accurate numbers, however I will make them up/round them to make the math easier for me and you. But 2.2 microseconds is the true figure.

There was a tunnel/tube (a bit like the hydro collider thing), it is a doughnut shape. It is, let's say, 2000 meters from one point, all the way around, back to that same point. I now need to clarify the Muon lifespan. Those 2.2 microseconds is a Muon at 'rest'.

Science Josh – 2012 Edition

Right, they put some Muons in the doughnut thing (don't know how, they would just 'die' after 2.2 microseconds but they did anyhow). And turned them (I think) into cosmic ray produced Muons. These Muons travelled at 98% the speed of light, that is: 293,796,608.84 meters per second.
Now here is the math, how many times should the Muons go around the doughnut before they 'die'?

Well:

293,796,608.84 = (s)

s/1,000,000 = Meters per microsecond (mpms)

mpms*2.2 = Meters travelled in life time of a Muon

That's is = 6463.52539448 meter in Muon lifetime

So, to work out how many times it should go around the doughnut before it 'dies' we need
to divide 6463.52539448 by 2000.

6463.52539448/2000 = 3.23176269724

When they did the experiment they had two counters/clocks; one outside the tunnel, observing (counting) how many times the Muon went around; and one inside, counting how many times the Muon passed a point.
The outside clock/counter recorded the Muons traveling around 3.23176269724 times (not actual number, the real

number was the same as their prediction based on similar math). However, the interesting thing is what the inside clock/counter recorded. It recorded the
Muons travelling around 16.1588134862 times.

Now, what does this mean? Was the clock/counter wrong?

The Muons, if you changed the math around, lived 5 times longer traveling at 98% the speed of light compared to their resting life time. That is, they lived 11 microseconds instead of 2.2 microseconds. This is important because:
In the tube/tunnel, to the Muons, 11 microseconds passed. But to the outside clock and people, only 2.2 seconds passed. So the Muons must be 8.8 microseconds ahead. Meaning that, it must be possible to advance things through time. Does that all make sense?
  The Muons travelled through time!

Science Josh – 2012 Edition

Time Travel: Back to the future
July 2012

The film; Back to the future, features a 'time machine car' which, when travelling at 88mph can travel through time.

Obviously this didn't actually happen in the film, but would it be possible to travel through time at 88mph? How far into the future could you travel?

Based on the maths and principles from my last post on time travel, I have calculated how far into the future the 'car' could travel.

I came to a conclusion that the car, traveling at 88mph or 36.33952m/s would advance in time by 0.000000000731810222 microseconds per second of travel.

In the film the car travels at 88mph for about 5 seconds before vanishing. So it would only be able to advance 0.00000000365905111 microseconds into the future. Considering that 1 microsecond is 1 millionth of a second, that figure is a tiny amount. It would not be possible, traveling at 88mph for the 5 seconds which feature in the film, to travel through time to a noticeable/measurable degree.

If you want to find out how I got those numbers, read on; if you don't care, then turn forward to the next post.

Science Josh – 2012 Edition

The math:
I used the formulae: (S/C x 100)xA
S= speed of travel (m/s)
C = Speed of light (m/s)
A= Advancement rate at 100% speed of light.
S and C are easy to calculate, the speed of light is: 293,796,608.84m/s.
S is the speed of travel, i.e. 88mph in 'Back to the future' it needs to be in m/s but this can be converted; 1 mph = 0.44704 m/s. 88mph is equal to 39.33952 m/s.
To calculate A it is slightly more confusing.

How A is calculated.

Using the information from my last post:
At 98% the speed of light time advances at 8.8 microseconds per 2.2microseconds
You can work out the rate at which time advances when traveling at 100% the speed of light:
(4/98)x100= 4.081632653061224
And then that multiplied by speed of travel as a percentage of speed of light. (SOL/speed)
That equals A= 0.00005465327526530612
And that is how you can calculate the time advancement of an object in relation to the speed of light.

Science Josh – 2012 Edition

Understanding the incomprehensible
July 2012

Today I propose the question:
Will the human race ever understand all that we crave to grasp?

Human beings, us, are yet to understand or fully grasp concepts we, ourselves, have invented. From the fourth dimension to the creation of the uncreated, will humans ever be able to understand what is currently incomprehensible?

Now, I have done no research relating to this however, I understand it is a controversial topic. I, myself, am a Christian, and therefore have a view that God is all knowing and has all authority over creation.

So, will we ever understand what a fourth dimension would look like?

Thanks by the way to Jasper Menzies for sending me a link to a YouTube video which inspired this bit:
That video can be viewed here on YouTube.

http://youtu.be/eGguwYPC32I

If you consider what a fourth dimension is, and regard it not to be time, but an inner view, then you realise the issue in understanding the ways of an entire universe that is four dimensional. For us to truly understand something in science we must observe it. The evidence for something is very important; just as mentioned previously about the universe expanding into nothing; and that nothing is: the absence of an evidence for the existence of something and not the absence of something. However if the case is, that to understand something, we must have evidence for it/observe it. Then we will never understand the fourth dimension, because, surely, we cannot observe the non-observable because, we see in 2/ 3 dimensions.
And in the same way, we can never therefore understand or think of creating something which incorporates nothing currently created. In other words, we could never even think of an animal that hasn't already been invented and has no body parts that already exist. If the Bible is correct, then that is what God did, when he created humans and all animals.

So this brings us back to my earlier assumption that something can only be understood of it can be observed. Before observing flight, would humans have been able to understand it?
No, humans would not be able to understand how flight works or is possible without seeing it.
That, therefore, brings us to a conclusion that we cannot understand the currently incomprehensible unless we observe it.

Science Josh – 2012 Edition

Global warming: Is it true?
August 2012

Is global warming true?

Well, first we need to find out what global warming is. According to Google it is: 'A gradual increase in the overall temperature of the earth's atmosphere generally attributed to the greenhouse effect caused by increased levels of carbon dioxide, CFCs, and other pollutants.'

So to find out if global warming is true, I will, over the next few chapters investigate the following:

1. Increasing levels of carbon dioxide:
      What does the CO_2 do?
       Is it causing global warming?
      What can we do?
2. Increasing levels of CFCs:
      What is CFCs?
      Is it causing global warming?
      What can we do?

Read on for an introduction to Global Warming.

Science Josh – 2012 Edition

Global warming: An introduction
August 2012

Here is a short introduction to global warming and the evidence for it happening:

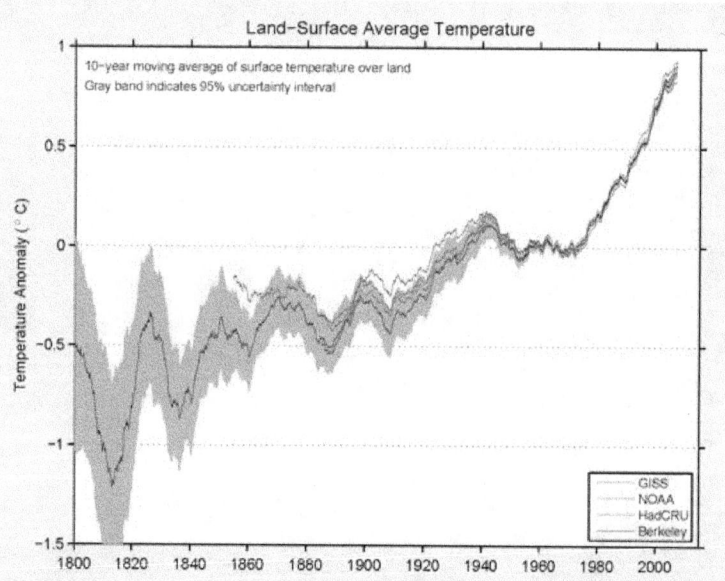

The graph show how the average temperature of the earth has increased from 1800 to 2000. The graph reads that an increase of 1.5°C has taken place in 200 years. That graph represents average land surface temperature. Now, it would not be wrong to assume, that that graph is evidence for global warming.

However, I would like to dig a little deeper into if that is the case!
Between 1800 and 2000 the world became more industrialised and farming became less common. As a result of all of the industrialisation and urbanisation, more ground was covered; concrete and tarmac, more houses were built and so on. If you then put that in context with how the average land surface temperature will change it gives some interesting results. It is, I think, simple to understand that open land like a field will have a lower land surface temperature than say covered land like a road. On a sunny day, the tarmac black road will have a higher temperature than the grass verge. The darker colours absorb and radiate more heat. So the reason for the increase in land surface temperature could be caused by urbanisation and industrialisation. But that is just my theory.

Here's a quick explanation of how the earth/atmosphere is warming up supposedly:
The heat enters our atmosphere from the sun. It passes into our atmosphere through the ozone layer (which doesn't have a hole in it). The heat travels through the atmosphere and reflects off the earth, warming the earth and our air. The heat then travels away from the earth and most of it should pass out, back through the ozone layer and into space. But apparently, the pollution (CO_2 etc.) is stopping the heat escaping, gradually raising the temperature due to the 'trapped' heat.

Science Josh – 2012 Edition

Global Warming: Carbon Dioxide
August 2012

This chapter is all about Carbon Dioxide and how it is contributing to global warming.

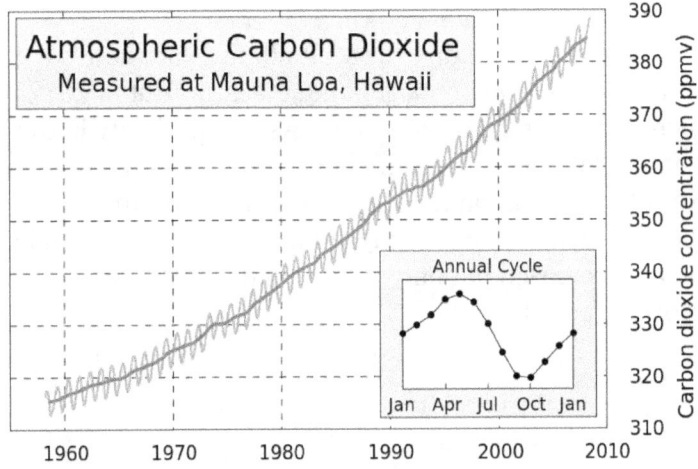

Yes, I am using yet another graph! So if you have a look at the graph you can see how the Atmospheric Carbon Dioxide levels are rising.
The Atmospheric Carbon Dioxide is basically the amount of carbon dioxide in the sky. Ignoring the annual cycle which is caused by seasons, the increase seems to be steady but steep. The levels seem to have rocketed from 1960 to 2010 by just over 70 ppmv. That sounds like a lot! If you don't know what ppmv stands

for, that does sound like a lot and would strike fear into every lover of polar bears. But if I say it, as it, what is means; it doesn't sound that bad at all. In the last 50 years the concentration of carbon dioxide in the atmosphere has risen by 70 parts per million by volume or 0.0000070 mg of carbon dioxide per 1 kg of atmosphere. That's a 0.0000007% increase in 50 years.

So now, you're thinking, that's tiny! Maybe Carbon dioxide isn't the main culprit.

But why has that figure risen? Well, in 1960 there were just over 3 billion humans living on earth. But in 2012 there are around 7 billion of us! As you probably know, humans take in oxygen and breathe out carbon dioxide. So if the population is increasing so are the carbon dioxide levels. Simple. Oh, and of course industry and so on has increased but only to meet the demand of a growing population!

Science Josh – 2012 Edition

Genetic Modification: Possibilities and Capabilities
August 2012

Last Sunday I was having a discussion with a friend of mine; Peter Barnes, our conversation lead to us talking about would it be possible to have super powers. I told him that you could take a characteristic from an animal and 'insert' it into a cell's nucleus and create a person, which had that characteristic. This seemed to blow his mind! He rapidly suggested multiple possibilities, and that got me thinking. What could you do? What are the possibilities of GM, could you give a person 'superpowers'?

46 Human Chromosomes

Science Josh – 2012 Edition

Before I start talking about the possibilities I must highlight one issue; it is illegal to genetically modify an embryo/human, so none of this is actually 'possible', although could be done.

A main superpower for Spiderman is his ability to produce 'spider thread' from his wrists. Could a human be genetically modified so he/she has the ability to do that?

Well, yes and no. Supposedly you could take the chromosome from a spider cell which contains the instructions for producing spider silk, and modify a humans DNA to include this characteristic.

This has been done with goats, which you can read about here:

http://news.bbc.co.uk/1/hi/sci/tech/889951.stm

The issue with using that technique to get the silk is that you cannot shoot it from your wrists. It is a protein, so you wouldn't be able to do anything like Spiderman. Sorry.

How about invincibility?

Well to do that we would need an invisible animal. Anything come to mind? No? Exactly, there is nothing we know of that is truly invisible. But there always could be invisible things that we have not discovered because we cannot see them!

Science Josh – 2012 Edition

Global Warming: CFC's
August 2012

What is CFCs?

'Chlorofluorocarbon or CFC is an organic compound that contains carbon, chlorine, and fluorine. The chlorine diminishes the effectiveness of the ozone layer and was used in refrigerators.'

Does it contribute to global warming?

Well, No. It is said to be a contributor but I cannot find out why. There is little of this present and the hole that was in the ozone layer which was caused by this is no longer an issue. So no, this seems to have no effect on global warming.

Why have you written about it then?

Science Josh – 2012 Edition

Is this life real?
September 2012

Is this life, that you and I are now living in real? Or is this the imaginings of a larger being or a computer simulation.

Well, what would a fake world look like? In a world that was imagined by something, would we have free will?

No. Surely we would not have any choice as to what to do, our destiny and lives would have been decided by the person who is imagining and; then automatically take place regardless of our imaginary preference. Also we would most likely not have a conscious, the ability to think, decipher between right and wrong and make informed complex decisions would probably not be possible if we were not real.
As philosopher René Descartes said "I think therefore I am".
The fact that we have a conscious and are able to make decisions means that we must be real.

But what would it mean for us if we were just a simulation?

Well, it would tell us the meaning of life. The purpose of the simulated existence is to be manipulated and to do as expected of the programmer. The thought that we could

be the imaginings of someone or being controlled by a being is shocking. But what would be controlling us? If we were just mere fictional characters who or what is the big being? That is an unanswerable question. At this point we are making many assumptions to get to this hypothetical position and as a result it is impossible to answer.
Could something be in-control?

Now we have showed we must be real, could there still be a higher being that is using us like chess pieces? No, again we have free will. Or could we all be playing the chess game ourselves with ourselves as pieces. The being is just the observer and is adjusting things before our move even. I am alluding to the question is there an 'adjusting' God or simply an only observing God?

Rewinding back to Descartes 'I think therefore I am'. This beautiful claim woks for one sole reason. If our lives were being imagined and controlled by another being; when we think: 'am I real?' Would not happen. Let me explain. As an imaginary person you hypothetically could think: 'Am I real?' But why would that happen? The imaginer already knows you are not real, he or she or it is imagining you. For the reason that the bigger being does not care if you think you are real or not. That would not happen. When Descartes said 'I think therefore I am.' That is what he meant!

Science Josh – 2012 Edition

What makes a human?
September 2012

What is a human?

Science has changed so quickly that it is now capable of some incredible things. In theory what I am about to explain is possible but has never been attempted in full. We have the ability to easily grow cells in a lab; they replicate themselves just like bacteria. These cells can then be 'moulded' into a desired shape. If these cells are specialized then the specialized cell would have to be bred from stem cells but the process would be the same. The collection of cells in the shape of say an arm could then have prosthetic bones and tubes for veins and arteries inserted into the shape. Receptors could be placed under the 'skin' and send electronic impulses up wires which then join with the real nerves in the shoulder. The arm could be fully connected up. Muscle tissue could be bred and inserted, connected to the nervous system and the correct neurons the muscles could be controlled by the brain. If this was practiced many many many times with lots of 'edits' the arm could become normal and usable just like the original. This sort of idea could be revolutionary for prosthetic limbs.

But how far should we go with this? We could, in theory, replace all your arms and legs when you get old. And, replace your heart with a fake one, your lungs with man-made lungs and your whole body. You would be as fit as a fiddle, everything would work like new. What a great idea! But is that the same person? The only thing that would be the same would be the brain cells. And even then, your brain slows as you age, so that wouldn't be as good.

Regardless of the moral reasoning's behind not changing a person's entire body and the issues with an ever aging brain. The concept of 'growing' body parts could change medicine for ever.
I certainly would not opt to have this done to me. The risks are tremendous and it would be impossible for the task to be completed including research in a lifetime anyway probably. But for a paralysed individual or someone with missing limbs; this could be great. Shame it won't be happening in the near future.

Science Josh – 2012 Edition

Video: Inside a black hole
September 2012

I was shown this by a friend, and although the guy in it is a bit over the top and slightly annoying, I found it very interesting. The concept that anything could become a black hole and how even light cannot escape the gravity is very interesting. Watch and enjoy!

Here's the link:

http://youtu.be/3pAnRKD4raY

VII. Inside a black hole

Science Josh – 2012 Edition

Welcome to the launch
September 2012

All has been changing on Science Josh so the following is from my opening post when I launched sciencejosh.com

Welcome to the launch of www.sciencejosh.com
 I am so glad you have arrived, have a drink, kick back and relax as you enjoy the wonders of science and my mind combined into a mixture of blog posts. Please look below at the previous post, visit the book shelf to view my latest eBook, released just yesterday!! At 5:35 UK time a new post should be released called Mind blowing. It will appear above this one. Make sure you don't miss it as it is a great read.

While your hear don't forget to follow me on Twitter: @gingerjoshprice and like the Facebook page: Science Josh
 Once you have done that please do share the site and my posts with the buttons below the posts. I hope you have a lovely time, any problems I will be on hand to help: contact@sciencejosh.com

All the best,

Josh Price

Science Josh – 2012 Edition

Mind Blowing
September 2012

Last night I settled down in bed and prepared to read myself a nice bed time story. The book in question: The Big Questions: Physics by Michael Brooks. This interesting and fairly easy to read book answers some of the lesser asked questions but remains a great read!

Anyway, I settled down and started reading; Can I change the universe with a single glance? Well guess what! You can! I was completely not expecting that! It turns out that in quantum theory some strange and weird and mind blowing stuff takes place. Let me tell you a bit about it.

Imagine this, a large sheet of paper. With another, just behind it, say a 30cm gap. In the front sheet of paper are two slits with 30cm between them. A single photon is fired at the paper.
Now let's think about what is going to happen. 1 photon, 2 holes, it is going to go through one gap. Yes?
Well, what happened was that it went through both, it left patterns on the paper behind, behind both gaps rather than the one which you would expect. So the experiment is reset and this time a detector is put on one of the gaps. The photon is fired and goes through one gap. What!?
 So the photon changed what it was doing? I can't think how 1 photon could go through both gaps but then when

observed travels through only one! Is the photon inconsistent, is there a pattern? No!

The photon when directly unobserved, acts like a wave and travels through both. But then seemingly changes state when observed? Mind blowing.
Then there is this very confusing thing called entanglement. It means that if you take two electrons, A and B and change the speed at which A is travelling, B will also change! These two unconnected electrons are somehow connected. This works from either side of the earth, change one and the other one changes.

Imagine cutting down a tree in one forest and another also falls as a result, but in a different forest all together!

"Einstein had spent the past two decades growing increasingly frustrated by the pioneers of quantum theory. Their ringleader Niels Bohr, claimed that the weirdness inherent in the theory, such as atoms existing in two places at once or effects preceding their cause, could only be explained if nothing – not even the moon – really existed until it was measured or observed." – Extract from the aforementioned book.

That puts the old, 'If a tree falls and no one is there to hear it fall, does it make a sound?' to complete pointlessness because Bohr says that unless someone is observing the tree, it does not really exist! I will come onto that later.

Science Josh – 2012 Edition

The New Magnesium
September 2012

Some of you may have seen the above image before; it is the new profile picture for the Facebook page and is part of our adverts. (Now the cover of this book)
But what is Magnesium and why have I chosen it to represent my blog?
Well Magnesium is a very abundant material, 9th in the entire known universe. And that's where I would hope my blog gets one of the top ten in the known universe.

Secondly, it is widely used in school and science labs across the UK and again that would be cool if my blog was used widely in schools.
And lastly it burns a brilliant white light! Check it out on YouTube! Rather Cool!

Science Josh – 2012 Edition

You are levitating right now!
September 2012

Yes! It's true! You are not touching the chair or the floor or well, anything! Below is a video from Vsauce which explains all of this!

http://youtu.be/yE8rkG9Dw4s

What does this mean?
If we can't touch anything apart from our parents and children, then what are the repercussions of this?
Could a 'murderer' say he didn't even touch the victim? Because technically he didn't.
Could you fall great distances because of course, you can't actually touch the ground? NO, of course not. But why? Well these repelling forces would be what you would feel. In fact you wouldn't have been injured or killed as a result of the fall, or hitting the ground but because the thing that would have killed you would be, your body not 'liking' the ground. Strange to think about it like that.
That knife didn't cut you; your skin just pushed against it and moved out the way leaving a gap.
Or, am I touching the air? Well, no, oxygen for example has 16 negative electrons which would repel/ be repelled by your body's atoms.
  You can't touch anything!

Science Josh – 2012 Edition

Very interesting facts!
September 2012

Quick explanation of powers:

A power is the little number often small and raised up by the bigger number. The web doesn't like that so instead I will have to use the ^ to indicate the next numbers are to the power of the number before it. 5^2 means 5×5. 5^{10} would be 5x5x5x5x5x5x5x5x5x5, or 5 with 10 zero's. Simple.

1. The probability of a planet that is conducive to life forming by chance is $10^{10,123}$. That's a planet that is suitable for life not that life is created.
2. The probability of life occurring by chance is $10^{340,000,000}$. To put that into perspective the are 10^{80} atoms in the known universe.
3. The chance of obtaining a single protein by chance combination is the same as filling the solar system with blind men, and them all solving a Rubik's cube simultaneously.
4. Imagine the entire earth filled entirely with peas, that is how many water molecules are in an orange.
5. DNA is 3.5 billion characters long. Each and every cell in your body apart from gametes contain that!
6. It is often thought that if enough time passes, life will eventually result. But it is less often noted that it would take a monkey 7.2 trillion trillion trillion trillion trillion trillion trillion years to type the first verse of Psalm 23.

Science Josh – 2012 Edition

Yet Richard Dawkins and so many others still believe that is how life formed, by chance!

All these facts are from various books and things I have seen over the years. I didn't work them out myself. Many are from Andrew Wilson's response to The God Delusion

Science Josh – 2012 Edition

Is Harry Potter Real?
October 2012

You may or may not have seen my Facebook post which related to this subject. I turned the tables and asked you what you think. Now I am going to try and construct a coherent response to this issue.
The 'fairy-tale theory' is this; all fictional characters that authors imagine and put into books are actually real and live in a parallel universe to us.
This idea has probably been around for ages however my English teacher raised it and claimed that was (hypothetically) her belief and her religion. Here is why she is wrong and the theory is not true.

1. 'Fairy-tale theory' is not a religion.
The main features of religion are that it gives hope, has morals and answers or explains things. Christianity for example gives hope; life after death and forgiveness through Jesus. Morals; the ten commandments and so on, answers or explains things; explains the origin of life and the universe, explains what happens after death. 'Fairy-tale theory' provides none of these; there is no hope or morals or questions answered because an individual believes that fictional characters are real.

2. If it is not a religion what is it?
'Fairy-tale theory' is a belief and, as the name suggests a theory. But what is a theory? Well, a theory is a collection of ideas which explain something, 'fairy-tale

theory' I suppose explains what happens to fictional characters once the story ends. But the issue is, once a different theory comes along which explains the same thing better, it takes the place of that theory. For example the theory that the world was flat was disproved by looking at it from space and finding it is round. The old theory was replaced by a newer one which had more evidence.

Considering there is no evidence to support 'fairy-tale theory' then even one piece of evidence to explain a counter idea would be sufficient to take its place. That one piece of evidence is the lack of evidence for 'fairy-tale theory'. The explanation of what happens to fictional characters once the story ends is: they never actually existed and so continue to not exist and don't 'live on'. The evidence?

Well, have you ever bumped into Willy Wonka on the Tube? No, so the fact that there is no evidence for the life of fictional characters is evidence that they don't exist.

So is that cleared up? Harry Potter is not real!

Science Josh – 2012 Edition

Flying Cars?
October 2012

Hyper Tech Propulsion

Flying cars, you hear about them rather often. But currently, they don't seem very popular. I haven't seen anyone just nipping to the shops in one.
Really expensive prototypes are about but you can't use

them on the road, or in the air. But why not? Some car manufacturer could make loads of money flogging them! It can't be too hard to make flying cars, can it?
Well there is a large issue with power. As you know if you have ever seen an aeroplane engine; they need a lot of thrust to get them in the air! And to fuel this they have large engines which power the motor at high speeds.
That engine runs on fuel, fuel has to therefore be carried on board. That make the plane heavier and so the motor has to work harder to get an equal amount of thrust. You clearly, in a small 4 person size car, couldn't have huge tanks for fuel or massive motors because there would not be the space.

So is that it, no flying cars?

Well no! Hyper Tech is the answer! The guys at HyparTech have created a design for a new propulsion device! This concept could power a flying car.
The design is on the previous page and below.

This device uses electromagnetic induced rotation to turn and channel the air through it. The device could revolutionise the propulsion market and the guys at HyparTech have done a smashing job. It works a bit like Archimedes' screw and was part inspired by him. Leonardo da Vinci then inspired the next part. He created this:

Science Josh – 2012 Edition

This works in a very similar way the Hyper Tech. It screws into the air just like a drill into wood. The issue was that he couldn't get enough lift to make it fly. The answer is in the Hyper Tech, it doubles the lift by adding a second channel. 'Yet also of primary importance is the cylinder that unsheathes the aerofoil. The cylinder is what allows for volumetric expansion & compression. The pressure differentials generated between the outside and inside through this volumetric expansion/contraction is what generates lift and temperature differentials.' The Hyper Tech can also achieve greater lift because it can spin at a greater rate than the above design.
HyparTech's great design is still in developmental stages however the concept has been proven to work.

How long until we see flying cars then? Well, flying cars may take a while, however as far as propelling boats or cars, this concept could take effect in our lifetime.

Science Josh – 2012 Edition

If a tree falls and there is no one around to hear it fall, does it make a sound?
October 2012

Some may try and respond with: Well that depends on what a sound is! They would be wrong to even ask that question. You see... The question we really need to ask, the fundamental question to answer the question of did that tree make a sound if no one was there is; If no one was there to see or hear it fall, how do we know it fell?

How do we know that tree even exists?

You may have just chuckled slightly; of course the tree fell and of course it exists: I can look at it now, lying on the ground!

Well, Quantum theory suggests otherwise. Due to entanglement (rather confusing) the tree does not exist unless observed/perceived/measured. This means that anything and everything does not exist unless someone is looking at it! So, if there is no one about to hear the tree fall, there is no one about to look at the tree and make it exist.

Let's stop there a second. Are you saying that if I sit in an empty room and close my eyes, the room ceases to exist and I just hover almost momentarily stuck, hovering in space time? Well, yes! And no.
Entanglement simply relates to relating. One atom can affect another one. They can change ones properties

as the other changes. Let's call our two atoms Bill and Ben. Bill is 'vibrating' at (hypothetically of course) 10vpm (vibrations per min). Ben is the vibrating at the same speed. If Bill is the atom in a football and Ben is in your foot. If Ben slows down 'loses energy' and now vibrates at 5vpm; Bill will also slow down to 5vpm! Think of that! 1 thing changes and another unconnected thing also does to the same degree.
Unfortunately I am only 15 and don't have a physics degree so am at this point out of my depth! But I will leave you with this:

http://youtu.be/1BfJ06plOTs

Einstein didn't believe anything I have just told you. He turned one day to a young physicist Abraham Pais and said: 'Do you really believe that the moon only exists when you are looking at it?'
In 1982 after Einstein's death, he was proved wrong. The moon only exists if you're looking at it.

Science Josh – 2012 Edition

The Higgs Boson
October 2012

Who hasn't heard of the Higgs Boson Particle or the God particle! But what is it, what does it mean and more importantly: Does it disprove the existence of a God?

Firstly, what is the Higgs Boson Particle?

The Higgs Boson is very confusing but here is a simple explanation of what it is:
Atoms are made up of subatomic particles: Quarks, Leptons and Bosons. An electron is a Lepton and a photon a Boson, for example.
Physicists ask, what gives subatomic particles mass? Or what is a subatomic particle made of?
Instead of saying 'just something even smaller – a sub-subatomic particle' they suggest this:
Everything 'sits in' a Higgs Boson field. Like a magnetic field or gravitational field. The Higgs Boson field is made up of 'particles' and these disrupt the subatomic particles. Just like a large object travelling through water will move slower than a small one. This disruption or 'slowing' of different amounts due to the resting mass of the sub atomic particles gives them mass.
That wasn't a mistake, the mass gives them mass. The idea works backwards sort of.
Anyway, all fields must have a source; a magnetic field has an electric current as a source, remove

that and the field disappears. The Higgs Boson particle is this source. However unlike their being billions of sources physicists suggest there are maybe 4 or 5! The concept is very similar to that of a thing called 'Ether' which made up the universe. It was like filing the universe with liquid called Ether and everything travelled through it. That theory was disproved in the 1900s.

Secondly, what does it mean?
For you and me? Nothing.
It means nothing and has no effect or real interest to anyone unless they want to know what gives subatomic particles their mass. It will advance quantum theory, however, will not revolutionise anything just yet as, as far as I am aware, there are no other theories resting upon its existence.

Finally, does it disprove the existence of God?

Well: If The Higgs Boson theory is proved that would mean that the Higgs Boson particle exists. And so consequently the existence of subatomic particles would be sustained by its existence. Atoms are made of subatomic particles and everything is made of atoms. So; you could say that the Higgs Boson particle sustains the existence of the universe.
Fair enough, right?
Well in the Bible God claims to sustain the universe and so we have a conflict.
God says he sustains the universe. And The Higgs Boson says it does.

You can either decide. Well that disproves God.
Or, like me say "God sustains the field and the Higgs Boson particle and controls the universe purposefully unlike the Higgs Boson."
But remember, this has not been proved yet and could end up just like Ether: disproved and useless.

Science Josh – 2012 Edition

The Higgs Boson – Take 2
October 2012

You may or may not have read about my post: The Higgs Boson.
Well, I received a comment from James Andrews (here's his blog: http://ursuscetacea.wordpress.com/). He highlighted some issues with my understanding in this field and kindly gave me some links which explain it well. I have watched them and am glad to report; I was wrong. My understanding was not complete and the information I provided was not 100% accurate.
I hope you will forgive me and to make it up to you, here are the video's that James kindly pointed me to. I hope these guys explain it better than me! Enjoy.

Part 1:

http://www.youtube.com/watch?v=9Uh5mTxRQcg&list=PLED25F943F8D6081C&index=13&feature=plcp

Part 2:
http://www.youtube.com/watch?v=ASRpIym_jFM&feature=BFa&list=PLED25F943F8D6081C

Part 3:
http://www.youtube.com/watch?v=6guXMfg88Z8&feature=BFa&list=PLED25F943F8D6081C

Science Josh – 2012 Edition

Dawkins
October 2012

Guess what book I got today! Yes, well done I got Richard Dawkins 'The God Delusion'.
Expect some of my future posts to revolve around his book! Should be an exciting,
interesting, controversial and belief-testing read!

Science Josh – 2012 Edition

Why do cats always land on their feet?
October 2012

Video time!

http://youtu.be/RtWbpyjJqrU

A cute, funny and informative video. I really recommend you watch this. Enjoy

Science Josh – 2012 Edition

Who made God?
November 2012

Who made God?

Nobody, God is eternal! But how on earth does that work? Well here is an extract from my upcoming book: God, Noah and the Big Bang. Enjoy!
"If God made everything he also made time. He made it and applied it to the universe which he also created. But if he made time then, before that point there was no time passing. Only now. An instantaneous, momentary, instant point. But with no end. It just repeats almost. A never ending cycle of instance.
As a result there was no beginning, no end and God is eternal.

But, why can't the big bang argue the same way? Well, if unassigned, unidentified matter had the ability and property of controlling the existence and creation of time; if that was something that matter could do (which it can't). Then that's God! Matter with all the controlling properties of God? That's just God.
God, I believe, created time and applied it to the universe but not to him or heaven. If this is the case it explains how eternity is possible. But also this concept reveals some incredible things. God sees, makes decisions, heals cancer, creates zebras and answers your prayers in an instant. That's how God can know everything about your future and you're past. He made you, answered your prayers and served you for your life time; watching you

grow up, caring for you in an instant. While sustaining the universe he did that!

You may be going, well that's rather impressive! I bet he's running about like mad: typing on a PC your answers to prayer, positioning the spots on a leopard and giving earth a little spin to keep it going. But guess what! The Bible says God is sitting on the throne! Sitting, reclined, having a drink of bubbly, relaxing. If I open Adobe Dreamweaver and loads of other programs at the same time on my computer, what happens?

They open obviously, it's a Mac. Jokes. It would freeze; the task manager would be hitting 80/90%!! But God? He stays on 0.0000001% maximum capacity. He sustains more than we can comprehend; but with such ease it is like flicking a light switch.

That, I think, is pretty darn amazing!"

Science Josh – 2012 Edition

The third law of thermodynamics!
November 2012

Wow, that' one heck of a title! The third law of thermodynamics!

So, what is it?
Well the third law states that absolute zero cannot be obtained.

What is absolute zero?

Well: Temperature is how much energy each particle has on average. The higher the temperature the higher the amount of energy per particle. Absolute zero is the temperature at which the particles have no energy. The third law states that this temperature can never be reached and, seeing we are yet to reach it. We should assume we can't and that the law is correct. But what if we could reach it?
What actually happens when a particle has no energy whatsoever?
Well, we don't know exactly. It is said that it would be in perfect crystal structure. Which isn't very interesting or exciting.

But I have a little idea which would be very interesting!

Science Josh – 2012 Edition

First though I suggest you watch this video and focus on WHY what happens happens:

http://youtu.be/eCMmmEEyOO0

I hope you enjoyed that little video, it really is interesting isn't it?
Now, if we imagine that slinky was at absolute zero. The bottom is being pulled down by gravity and up by tension. It balances it out. In the video the gentleman says that a compression wave has to travel down the slinky to let the other end know that the top has been released. A compression wave however is a transfer of energy through a substance, in this case a slinky but if our slinky has no energy it can't produce that compression wave.

What does this mean?

In theory the slinky would just remain still, frozen in mid-air if it had no energy.
Really! Stuff just hovers at absolute zero?

Well, no. With no energy, yes it would hover. But it would be impossible for it to have no energy the moment you held it above the ground you give it gravitational potential energy. It has energy again and so can fall. But still, that's pretty cool.

Science Josh – 2012 Edition

Stimuli-responsive polymers And Invisibility
November 2012

Sorry, what?
Stimuli-responsive polymers.
What on earth is that?
Well, they're also known as Smart Polymers, you learn about them at GCSE Level. And, they are very useful.

What are they?

Well they are polymers (large chains of a substance) which adapt or change as a result of their surroundings. These polymers react to changes in temperature, PH, light intensity and other various factors. They then can change shape, colour or transparency or contract in size. Smart Polymers are used
in medicine for degradable stitches and other bits and bobs.

But today I am interested in not being seen! I want to use smart Polymers to make myself invisible ish.
How?
To create this invisibility cloak I first need to decide on what polymers to use. I want the properties so that the polymer changes colour to match my surroundings.

So part 1: Get a Smart Polymer which changes colour.

But we need a stimulus. Now here is where it will get a little tricky. We want a very
sensitive polymer which reacts to temperature.
How will that work? Are you going to stand in front of a radiator to become invisible? No.
Remember the light spectrum?

Well each band of light has
a slightly different temperature: Red is hotter than Blue.
As a result when white light is shined at a red surface
only red is reflected, all other colours are
absorbed. This red will be slightly higher
in temperature than, say, a blue reflection.

So part 2: Get a polymer which is stimulated
by temperature.

And boom! You have a polymer, which, once you cover your body in it will make you invisible!

*I must point out, this most likely wouldn't work. But it's a nice idea anyhow!

Science Josh – 2012 Edition

Competitive Pain – The Science of Pain
November 2012

We can remember that type when we got a paper cut and it hurt, right?
But, unlike other emotions and feelings, you
cannot remember or reproduce that explicit feeling of pain.
Take a cross-country run. 1.5 miles. You finish in 6 mins 32 seconds. You're gutted; your target was under 6 mins. You think to yourself on your way home "I wish I had pushed myself more, I know I could have."
But when you were running and in
pain from exhaustion and sore feet and your face feel red-raw from the cold winters wind. That pain was bad. You wanted, in your mind to push harder but you were held back by this feeling that you were in danger or in too much pain, that it might be bad if you go faster.
That is what I am interested in for this post. I was asked by a Twitter follower (@JoshSamworth) to do a post about Neuroscience. That's not my area of expertise so I was a little stuck for what to write about. But after half an hour on Google and Wikipedia I came up with this post. I want to investigate pain and its effects
on memory and how this limits our human capabilities. If we could overcome or understand how pain affects this we may be able to push past our natural pain thresholds or barriers.

So, what is pain?

Science Josh – 2012 Edition

Pain is physical discomfort or suffering. And, although physical is actually not real. Our body tells itself to be in pain as a reaction to something. When you stand on something sharp like a pin your skin receptors pick it up and tell the rest of your body that there is lots of pressure on that part of the skin. You feel pain and your motor neurones tell muscles to take your foot away. Pin, although often annoying is mostly
very useful; it stops us burning ourselves and d protects us from serious injury. Some people cannot feel pain and often die rather young after injuring themselves without noticing.

Why would our brains stop us remembering pain?

Well, of course no one would want to be able
to remember only unpleasant things so, your brain forgets them. Please often say how good the 80's were and so on. That's because they can
only remember mainly the positive bits and
cannot remember how they cut their figure on a knife and stubbed their toe on the bottom step ten billion times. So our brain ha every benefit to forget how it felt to be in pain. But we still remember that it hurt and so we don't do it again. The brain teaches itself almost. After some research I found some useful reasons for experiencing pain while running:
Firstly, if you are running a long distance you get into your rhythm and a good runner could go on for ages. While doing this physical exercise your blood is pumping fast around your body which means you may

become light headed or dizzy and pass out. But your body puts you in pain; this keeps you alert and stops you passing out.
Secondly, intense pain can do some incredible things to your body such as shut-down (coma) or go into shock and so on. But sometimes it can have an opposite effect. It can give you an adrenaline rush. The more intense pain puts you in a fight or flight type response. This allows you to run even faster and is a natural aesthetic for the pain. It allows you to concentrate more; some people feel a slow motion type effect. All of this allows you to stay at the top of your game and is a natural instinct to help protect from predictor or dangerous situations.

How can we harness adrenaline?

Adrenaline gives us a boost and allows us to push our bodies further than normal activity. There are many ways that you can give yourself an increased adrenaline boost. You could, while on your 1.5 mile run run on top of homes and houses or at a great height. This is dangerous and so your body will release adrenalin to try and keep you safe. However, that most likely isn't for everyone. The other thing you could do is be chased. Get someone to chase you and you will run faster and feel a sense of danger. Your adrenal glands
will excrete adrenaline and boom; you have pushed past your natural hum. Or, the easier option is not to jug but to run. The faster you run the more likely that you will feel in danger. And, your body will help you along.
That's one way we can control our pain but overall, pain is positive and temporary as we forget it almost instantly.

Science Josh – 2012 Edition

This is a bit long winded but shows what I mean nicely:

http://youtu.be/4g25d7_Afmc

Science Josh – 2012 Edition

Origin: The creation of the universe, the laws of science and time
November 2012

One of the largest perceived differences between Religion and Science is the creation of the universe. The origin of the stars, moons, suns and hydrogen gas is a much debated issue. I hope you enjoy my following thoughts!
The Big Bang suggests that the universe started with, almost nothing more than, a single point in time and space. At this single point was highly condensed matter; very, very hot!
It expanded from that point rapidly – Not exploded; there was no oxygen meaning no combustion!

As the universe expanded approximately 13.75 billion years ago it cooled. This cooling allowed energy to be converted to subatomic particles like electrons and protons; the building blocks of everything. After a few minutes neutrons and protons joined creating positively charge particles. The dense mass expanded and cooled and after a few thousand years electrons joined the neutrons and protons creating neutrally charged particles – atoms. The first element was produced – Hydrogen and small quantities of Helium and Lithium. The universe continued to expand and these elements were drawn together due to gravity. These growing masses within the three-dimensional expansion of matter and energy formed what we know today as stars and galaxies.

The matter continued to expand and heavier and heavier elements were formed. These heavy elements joined together and, like a snow ball in snow, grew to form planets. These planets were attracted to the stars which were now much larger than them. These planets started orbiting the stars and this happened thousands and thousands of times over. Until, we get to modern day; it is predicted that there could be over a billion billion planets spread over the twenty-six billion light year wide universe.

That's a brief description of how the Big Bang concept works. It's widely accepted and, many Christians say that is what happened – that was planned by God. I disagree with aspects of the theory.

Although I do not advocate a 'God of the gaps' idea it is important to point out the following issues: The Big Bang theory does not explain the origin of the matter and energy – frankly this renders the theory useless for the explanation of the universes origin. Secondly, at that initial point there was lots of highly condensed mass at high temperatures – but what was that made from? If protons and electrons could not form because it was too hot we have to ask: what was it made of? Thirdly, where did all those laws and rules which physics is based on come from? Planets were formed because gravity pulled elements together but where did that gravity come from? What is there to say that X, Y and Z must happen and apply to A, B and C?

You may think I am just being picky, you may say: "Well, they have just always been there."

Are you serious? Your saying that those (unidentified) particles have just been sitting there forever and then one day – POP – and the universe is born. Where did time even come from? What made time? 'It's always been there' without a logical answer as to how that is possible does not answer the question of the origin of the universe. The big bang theory is redundant by that logic. But, let's not abandon it there. The big bang produced planets. These planets, like us, circle a star.

They whiz around the outside of a very hot ball of gas. The temperatures are so hot in these stars that elements undergo nuclear fusion. Our planet, by chance; according to the big bang theory, was just in the right spot. We are in what's called the goldilocks zone. Where it's not too hot, but not too cold. You may think, "Okay, that is fair enough, by chance we ended up in the right zone." But there are loads of 'zones'. We have water because of our temperature but we also have an atmosphere and the correct elements were on earth for abiogenesis and there was a moon to control the tides and Jupiter was near enough to pull away meteorites which could have hit the planet. We were in the dark for only 24 hours and the sun always shined on us in the day.

Imagine if we had no water, it took the earth 150 hours to spin on its axis. Plants would not survive. Humans would not be able to stay alive.

Here's some numbers to put this into perspective:

The chance of a planet being created by chance which is conducive to life is: 1 in $10^{10,123}$ That is a lot! Just so you know how much that is, there are 1080 atoms in the entire known universe!

So, you're probably wondering: "What do you think then?" "What's your amazing theory which solves all of these issues?" "God!?"

Well, yes! God!
I believe this:
God created matter, energy, time and gravity. He poured them into his universe like an egg yolk into a cake mixture. They spread, filled out. God kept on pouring the energy and matter into the universe. The gravity God had made pulled the matter he had made together, just as God had planned. Planets were formed. The Big Bang is very similar to this after that initial point.
I love the phrase: I believe in the Big Bang! When God made everything I am sure there was a pretty big bang! But, you may now say: "Ah, but who made your God?" This is often asked and answered by people with little understanding in this. Many say, like me, God has always existed, God is eternal. That is useless, the big bang says its matter has always existed – It is eternal. If God is eternal, why can't our tiny condensed matter be also?
'The eternal God' argument without explanation is useless, but with explanation it's incredible!

Science Josh – 2012 Edition

Why are my headphones tangled, again?
December 2012

So annoying isn't it? You want to listen to some Coldplay or Heavy Metal and you take your headphones from your pocket and...They're tangled up. You then spend so long untangling you run out of time to listen to your tunes.
But, why do they get tangled? They enter in lovely and coiled, tangle free. But 5 mins later they are tangled! Well, it's all about probability. Your headphones went into your pocket untangled. There is only one way they can be not tangled, how they are at the moment. But there are hundreds of ways that they can be tangled and knotted.
So if you put them into your pocket untangled there is only a 1 in 800 chance your headphones can come out untangled and an almost 100% guarantee they will come out untangled.
This is called knot theory. In mathematics it refers to circular knots, string circles with no end. But your headphones have ends, two. But the principle still applies.
This concept works in many different situations, your bedroom will get messier and messier because it is only 'tidy' in 3 or 4 ways but there are 1000's of ways it could be messy. Your car is more likely to break than work. There is 1 way that everything works in your car but many things that would make it not work.

I could end this post here, but that would make it too short. So instead I am going to do a little experiment! A first for Science Josh!
I am going to try to solve all headphone tangling issues!

Here is how it will work:
1. Take a pair of headphones, a normal pair of headphones
2. Take a timer and place the untangled headphones in your right front trouser pocket, start the timer
3. Walk around for time intervals before removing the headphones and recording the knotting pattern.
4. Repeat the experiment with headphones in back pocket, wrapped around a phone, tied in a purposeful knot
Here are the results:
There was a clear correlation between the number of knots and the time the headphones were in the pocket. I found a 1:1 correlation where if you double the time you double the number of knots.
When the headphones were wrapped around the phone they came out untangled but when removed from the phone they became knotted.
When tied in a purposeful knot the headphones came out tied in the same knot. This means that the knots can be easily untangled as you know how it is knotted.
In conclusion,
It is best to tie you headphones in a knot before putting them in your pocket.

Science Josh – 2012 Edition

I used the following knot:

http://youtu.be/Wfzt4jSOV7k

Science Josh – 2012 Edition

Could there be life on other planets?
December 2012

So, straight to the point! I was inspired
by watching Science Club on BBC. I watched last weeks
on iPlayer and felt it a good idea to discuss my thoughts.
Could there be life on other planets?
Well, Yes! I feel it is quite possible that we could find
life on the moon or even Mars. Why? Because we have
been there! Life can easily originate from life. By
visiting the moon we have almost delivered life to it.
That flag they put on the moon? That
was probably covered in thousands of bacteria. The rate
at which the reproduce means that mutation could occur
to allow them to survive. However, we
won't find animals or birds or fish
or intelligent creatures. So, really we need to rewrite that
question like this:
Could life have arisen on other planets? And if so, Can
we find it?
I would say: No
Some Scientists would Say: Yes
But some would join me on saying no.
Of course we cannot be 100% sure that life hasn't arisen
on more than earth (assuming life
did spontaneously arise) but I am 99% sure.
Firstly I must make clear I am
a Christian and therefore believe God designed humans

and placed them on earth to rule over the animals. I don't believe abiogenesis, that's the arising of life by chance due to a primeval soup, is possible.
I base my beliefs on that matter on probability. And use it as my partial evidence. I will for this argument also.
You see, if spontaneous arising of life is possible it is extremely unlikely. And so for it to happen more than once is twice as extremely unlikely. Still assuming that happened, the chance of the single celled organism evolving into anything like a human (assuming you believe we evolved from a single cell organism) is also extremely unlikely.
Let's look at some figures:
Chance of a planet being conducive to life:

or 1 in 10 to the power of $10^{10,123}$ Or 1 with 10123 zero's

Science Josh – 2012 Edition

Chance of two planets being conducive to life:
1 in 10 to the power of 20246 Or 1 with 20246 zero's

Chance of life arising:
1 in 10 to the power of 340,000,000 Or 1 with 340,000,000 zero's

Chance of life arising on two planets:
1 in 10 to the power of 680,000,000 Or 1 with 680,000,000 zero's

Number of estimated planets:
5 to the power of 32 Or 5 with 32 zero's

Chance we will communicate with said planets:
Seeing we only now of about 200 -about 1 in 5 to the power of 30 Or 5 with 30 zero's

Overall probability of life arising on two planets and humans finding that life:
About 1 in 10 to the power of 700,000,000.

So, considering there are only 500,000,000,000,000,000,000,000,000,000 planets we would need 218,750,00 times more planets for two to have basic life forms! Not even complicated ones like ours.

So, if you still believe that it is possible for life to arise on other planets, you are nuts.

All the figures came from different, trustworthy, unbiased sources. The 340,000,000 one came from an evolutionary scientist!

Science Josh – 2012 Edition

Will the world end this Friday?
December 2012

Many have said the world will be ending this Friday (21st December)
The idea is that in 2003 a planet was predicted to collide with earth. It didn't and was re-predicted to collide in 2012. This was then tied in with the end of a quarter of the Mayan calendar. This collision between two planets would wipe us all out and it would b the 'End Of The World'. But, lucky for you, I don't think that will happen. NASA say, there is no sign of any threats to our planet which could cause it to end which are above their normal possibility.
But here's why I know the world won't end. When we say 'the end of the world' we literally mean, the world stops being. But what we actually picture or contemplate is that everything stops. Everything stops existing. No one, as far as I know, goes: "Yes the world will explode tomorrow, and we will all die. But life will arise on the moon. Don't worry, life will continue due to the spreading out of our atmosphere and the substances on the earth". You see, if the world collided with another planet, life would, most likely continue. Earthly stuff would be spewed out all over the universe. This would be 'picked up' by the other planets and eventually added onto their surface. But our planet has all of the conditions for life, right? Plus, bacteria and that sort of stuff would most likely survive a collision between two planets.

Science Josh – 2012 Edition

So, what we actually mean with: ' The End Of The World' is: the end of time. The ending of existence. All will cease.
So, is that going to happen?
No.
Why?
Picture Christmas day, 2012. Got it?
You just imagined and understood the future.
If time stops on Friday, time doesn't exist for the 25th to take place. I believe, if time was going to stop on Friday, that we would not be able to comprehend time past that point. Why? Because we cannot comprehend the non-existent/non-existed/non-to-exist.
Can't think of a new way of breathing? Can you? You can't make up stuff that hasn't or isn't or isn't going to exist.
And that's why time won't stop, life won't cease and why the 22nd will be cold and rainy. :(

Science Josh – 2012 Edition

#2 Will the world end tomorrow? – From the Bible
December 2012

It is the 20th of December, 1 day before the world is said to end. Already I have addressed the scientific reasoning behind is's ending and concluded it won't end. You can see some of my debates on this on the Facebook page: www.facebook.com/sciencejosh . Now, I want to look at what the Bible says about the end of the world.

Some may doubt the reliability of the Bible. But I believe it's governing author (God), the creator of the universe, will be 100% correct about the ending. He is the one ending it after all.

So,

Will the world end tomorrow?

Section 1: How does the Bible say it will end?

I won't quote the Bible much as you can just read the section on this in the Bible; Matthew 24 vs 25 onwards .Instead I will paraphrase mostly. The Bible starts with Jesus saying that many will pretend to be the Messiah. The Bible says that Jesus will return and earth and heaven will become one and judgement day will happen and there will be believers going to Heaven and non-believers to Hell. Jesus says, many will pretend to be me, don't believe them. Why? How do we know who is and isn't Jesus? The Bible says this: "And then all the peoples of the earth will mourn when they see the Son of Man coming on the clouds of heaven, with power and great glory. And he will send his angels with a loud

trumpet call, and they will gather his elect from the four winds, from one end of the heavens to the other."
That pretty much sums up step 1, don't mistake anyone for Jesus; peoples of the earth will surely know when he has come.
But before Jesus comes the Bible says the sky will darken, the moon will not shine and the stars will fall from he sky. (That has not happened!)
Earlier in this section however, it says there will be earthquakes. (We have had those So, is this a sign the world will end? There have been earthquakes, people have died and it has destroyed some things. But when the world is ending, I am sure that the earthquakes will be more than buildings falling down in Haiti. The WHOLE earth will feel the powerfullest shaking; for the wrath of God will be moving.

Section 2: When will it end?

The Bible says a sentence which tells me 100% that the world will not end tomorrow. It says: "But about that day or hour no one knows, not even the angels in heaven, nor the Son,". Whoa! The Bible says we will not know the time or date. Not even the angles, the holy servants in heaven will be told. Then, it says, "nor the son"!!! Not even Jesus will know when the world will end! Not even Jesus!

Conclusion

The Bible clearly says that not only will not even Jesus know when it will end, let alone us, but that it will be powerful! You will know about it. My Spanish lesson, your Christmas shopping, walking the dog; it will all be insignificant and unimportant because Jesus will be coming on the clouds of heaven!

About The Author

Josh Price is a 15 year old with a keen interest in science. He lives in Sussex in the UK and writes and publishes posts in his spare time. He is specifically interested in the universe and the Evolution vs. Creationism argument. He is a Christian who believes that God is the logical answer to the origin of the universe and life. He enjoys reading, researching and debating over science related topics. He is in his final year of secondary school and is studying, among other things, Science, Drama, IT and Geography. Josh's dream is that this blog will become successful and he will be able to express and explain his views and ideas to a wider audience. He firmly believes that in a few years he will be debating the likes of Richard Dawkins over the origin of life and the universe. He has an interest in cycling and voluntary work and looks forward to further education and work. Josh is working hard to make this blog a success on a tiny budget. It would be great if you could give him a hand by sharing it with your friends.

About the blog:

The blog was started in June as a result of boredom. I decided to open a blog and post about whatever was going on. I created the blog and was wondering what to blog about. I remembered a discussion I had had with a friend of mine over what the universe was expanding into. I decided this would be my first post. I wrote about it, it's called an infinite universe, and it was instantly popular. I was happy to see a fair number of visits and a few comments on the post. Recently I have seen great growth in visitor numbers and the blogs popularity and hope to see this continue. Please consider sharing the blog with everyone you know and joining me on social networking sites.

Science Josh – 2012 Edition

Thank you so much for reading this book.
This is the 2012 Edition so look out for future editions with more recent post's in them.

Please do continue to visit www.sciencejosh.com

Follow Us on Twitter: www.twitter.com/sciencejosh

Like Us on Facebook: www.facebook.com/sciencejosh

Or personally email me: joshprice01@gmail.com

Thank you again,

Josh Price
Scientist, Blogger and Author.

Science Josh – 2012 Edition

2012 Edition (Comprised Dec 2012)

www.ingramcontent.com/pod-product-compliance
Lightning Source LLC
Chambersburg PA
CBHW072210170526
45158CB00002BA/533